Solve for x $\sqrt{x} + \sqrt{x} - 4 = 10$

Solve for x $2^{(-x-2)} = \frac{1}{32}$

Simplify $\frac{4x - 12}{x^2 + x - 12}$

factor $\frac{4(x-3)}{(x+4)(x-3)}$

cancel $\frac{4}{(x+4)} \; \frac{(x-3)}{(x-3)}$

$\frac{4}{(x+4)}$

Solve for x $11^{4x} = 9$

using logarithms $log11^{4x} = log9$

$4x\,log11 = log9$

isolate 4x $4x = \frac{log9}{log11}$

isolate x $x = \frac{1}{4} \cdot \frac{log9}{log11}$

$x = 0.23$

Hard Math

Hard Math Problems
1st Edition

Think you can solve some of the toughest math problems?

Hard Math gives some some of the most challenging math problems. Though you don't have to be a genius to solve these, they are quite difficult and are the most likely for people to get wrong on a test. [1]

Hard Math is an excellent way to practice or improve your skills. It also provides a form of challenging entertainment. Step-by-step solutions are provided for each problem. If you're stuck, just study the solution.

[1] Depending on your abilities, you may find several of these problems to be easy while other problems are difficult. This book provides enough variation to practice many math and algebra techniques. You may need a calculator.

Disclaimer: The author makes no guarantees as to the accuracy of the material included. This book is for entertainment purposes only. The included solutions are only that of which the author would perform. There may be several solution methods and/or answers to the equations and/or expressions included in this packet. .

See pages 40-43 for math rules.

Copyright © 2018 Timothy Schablin

All rights reserved.

ISBN: 978-1720900375

Solve for x $\sqrt{x} + \sqrt{x} - 4 = 10$

isolate radicals $\sqrt{x} + \sqrt{x} = 10 + 4$

$2\sqrt{x} = 14$

$\sqrt{x} = \frac{14}{2}$

$\sqrt{x} = 7$

square both sides $(\sqrt{x})^2 = 7^2$

$x = 49$

Solve for x $2^{(-x - 2)} = \frac{1}{32}$

using logarithms $(-x - 2)log2 = log\frac{1}{32}$

$-x - 2 = \frac{log0.03125}{log2}$

isolate -x $-x - 2 = -5$

$-1x = -5 + 2$

$-1x = -3$

$x = \frac{-3}{-1}$

$x = 3$

Simplify $$\frac{4x - 12}{x^2 + x - 12}$$

Solve for x $$11^{4x} = 9$$

Simplify $\frac{x+1}{x^3 - x}$

factor $\frac{x+1}{x(x-1)(x+1)}$

cancel $\frac{1}{x(x-1)} \frac{(x+1)}{(x+1)}$

$\frac{1}{x^2 - x}$

Find f(x + h) $f(x) = 5 - 7x^2$

insert (x + h) $5 - 7(x + h)^2$

$5 - 7(x + h)(x + h)$

using FOIL $5 - 7(x^2 + 2xh + h^2)$

multiply $5 - 7x^2 - 14xh - 7h^2$

$f(x + h) = 5 - 7x^2 - 14xh - 7h^2$

Simplify $\dfrac{x+1}{x^3 - x}$

Find f(x + h) $f(x) = 5 - 7x^2$

Solve for b when $y = f(g(x))$ passing through point (4, 6)

$f(x) = \sqrt{x}$

$g(x) = 7x + b$

---

insert (4) to g(x)

$7(4) + b$

$28 + b$

insert (28 + b) to f(x)

$\sqrt{28 + b}$

solve for b

$6 = \sqrt{28 + b}$

$36 = 28 + b$

$36 - 28 = b$

$8 = b$

$b = 8$

Simplify

$3\sqrt{27} + \sqrt{12}$

---

factor each

$(3 \cdot \sqrt{9} \cdot \sqrt{3}) + (\sqrt{4} \cdot \sqrt{3})$

$(3 \cdot 3 \cdot \sqrt{3}) + (2 \cdot \sqrt{3})$

add coefficients

$9\sqrt{3} + 2\sqrt{3}$

$11\sqrt{3}$

Solve for b when y = f(g(x)) passing through point (4, 6)

$f(x) = \sqrt{x}$
$g(x) = 7x + b$

Simplify

$3\sqrt{27} + \sqrt{12}$

Solve for x $\sqrt{3x + 4} + 5 = 13 - x$

isolate the radical $\sqrt{3x + 4} = 13 - x - 5$

$\sqrt{3x + 4} = 8 - x$

square both sides $(\sqrt{3x + 4})^2 = (8 - x)^2$

$3x + 4 = x^2 - 16x + 64$

merge to equal '0' $3x + 4 - x^2 + 16x - 64 = 0$

combine like terms $-x^2 + 19x - 60 = 0$

factor $(-x + 4)(x - 15) = 0$

$(-4 + 4)(15 - 15) = 0$

$x = 4$ *or* $x = 15$

only x = 4 works in original equation

Simplify $(\sqrt{x} + \sqrt{2})^2$

$(\sqrt{x} + \sqrt{2})(\sqrt{x} + \sqrt{2})$

using FOIL $\sqrt{x}^2 + \sqrt{2x} + \sqrt{2x} + \sqrt{4}$

combine like terms $\sqrt{x}^2 + 2\sqrt{2x} + \sqrt{4}$

$x + 2\sqrt{2x} + 2$

Solve for x $\sqrt{3x + 4} + 5 = 13 - x$

Simplify $(\sqrt{x} + \sqrt{2})^2$

Solve for x $\sqrt{x - 9} + \sqrt{x} = 9$

isolate $\sqrt{x-9}$ radical $\sqrt{x - 9} = 9 - \sqrt{x}$

square both sides $(\sqrt{x - 9})^2 = (9 - \sqrt{x})^2$

$x - 9 = (9 - \sqrt{x})(9 - \sqrt{x})$

$x - 9 = 81 - 9\sqrt{x} - 9\sqrt{x} + x$

combine like terms $x - 9 = 81 - 18\sqrt{x} + x$

merge to equal '0 $0 = 81 - 18\sqrt{x} + x - x + 9$

combine like terms $0 = 90 - 18\sqrt{x}$

get radical on left $18\sqrt{x} = 90$

isolate radical $\sqrt{x} = \frac{90}{18}$

$\sqrt{x} = 5$

square both sides $(\sqrt{x})^2 = 5^2$

x = 25

Solve for x $\sqrt{x-9}+\sqrt{x} = 9$

Simplify $$\frac{5}{x+2} + \frac{1}{2}$$

---

find LCD $$\frac{(5)(2)}{2(x+2)} + \frac{(1)(x+2)}{2(x+2)}$$

$$\frac{10}{2x+4} + \frac{x+2}{2x+4}$$

add numerators $$\frac{10+x+2}{2x+4}$$

$$\frac{x+12}{2x+4}$$

Simplify $$\frac{x - \frac{3}{x-2}}{1 - \frac{1}{x-2}}$$

---

multiply by (x - 2) $$\frac{x(x-2) - \frac{3}{x-2} \cdot \frac{x-2}{1}}{1(x-2) - \frac{1}{x-2} \cdot \frac{x-2}{1}}$$

$$\frac{x^2-2x-3}{x-2-1}$$

factor $$\frac{(x+1)(x-3)}{x-3}$$

cancel (x - 3) $$\frac{(x+1)}{1} \quad \frac{(x-3)}{x-3}$$

x + 1

Simplify $$\frac{5}{x+2} + \frac{1}{2}$$

Simplify $$\frac{x - \frac{3}{x-2}}{1 - \frac{1}{x-2}}$$

Solve for x $\frac{3}{4}x - x = x - 5$

$$\frac{3}{4}x - x - x = -5$$

combine like terms $\frac{3}{4}x - 2x = -5$

multiply by by 4 $(4)\frac{3}{4}x - (4)2x = (4)-5$

$$3x - 8x = -20$$

combine like terms $-5x = -20$

divide by -5 $x = \frac{-20}{-5}$

$$x= 4$$

Find f ∘ f $f(x) = x^2 - 8$

insert f(x) $(x^2 - 8)^2 - 8$

$$(x^2 - 8)(x^2 - 8) - 8$$

using FOIL $x^4 - 8x^2 - 8x^2 + 64 - 8$

combine like terms $x^4 - 16x^2 + 56$

$$f(x) = x^4 - 16x^2 + 56$$

Solve for x $\frac{3}{4}x - x = x - 5$

Find $f \circ f$ $f(x) = x^2 - 8$

Solve for x	$\sqrt{x + 5} = 1 + \sqrt{x}$
square both sides	$(\sqrt{x + 5})^2 = (1 + \sqrt{x})^2$
	$x + 5 = 1 + 1\sqrt{x} + 1\sqrt{x} + \sqrt{x^2}$
combine like terms	$x + 5 = 1 + 2\sqrt{x} + x$
move radical to left	$x + 5 - 2\sqrt{x} = 1 + x$
isolate radical	$-2\sqrt{x} = 1 + x - x - 5$
combine like terms	$-2\sqrt{x} = -4$
divide	$\sqrt{x} = \frac{-4}{-2}$
	$\sqrt{x} = 2$
square both sides	$\sqrt{x}^2 = 2^2$
	$x = 4$

Solve for x $\sqrt{x + 5} = 1 + \sqrt{x}$

Simplify	$(x + 5)(3 - x)(-x - 2)$
using FOIL	$(x + 5)(-3x - 6 + x^2 + 2x)$
combine like terms	$(x + 5)(-x - 6 + x^2)$
using FOIL	$-x^2 - 6x + x^3 - 5x - 30 + 5x^2$
rewrite	$x^3 - x^2 + 5x^2 - 6x - 5x - 30$
combine like terms	$x^3 + 4x^2 - 11x - 30$

Solve for x	$-0.25x^2 + 1.5 = -10.75$
isolate $-0.25x^2$	$-0.25x^2 = -10.75 - 1.5$
	$-0.25x^2 = -12.25$
divide	$x^2 = \frac{-12.25}{-0.25}$
	$x^2 = 49$
sqrt both sides	$\sqrt{x^2} = \sqrt{49}$
	$x = 7$ *or* $x = -7$

Simplify $(x + 5)(3 - x)(-x - 2)$

Solve for x $-0.25x^2 + 1.5 = -10.75$

Solve for x $5 + \sqrt{\frac{1}{2}x + 9} = 12$

isolate radical $\sqrt{\frac{1}{2}x + 9} = 12 - 5$

square both sides $(\sqrt{\frac{1}{2}x + 9})^2 = 7^2$

$\frac{1}{2}x + 9 = 49$

$\frac{1}{2}x = 49 - 9$

$\frac{1}{2}x = 40$

multiply by 2 $(2)\frac{1}{2}x = (2)40$

$x = 80$

Find f ∘ g(x)

$f(x) = \sqrt{2x + x}$

$g(x) = 3x - 5$

insert g(x) $\sqrt{2(3x - 5) + 3x - 5}$

multiply 2(3x - 5) $\sqrt{6x - 10 + 3x - 5}$

combine like terms $\sqrt{9x - 15}$

$f \circ g(x) = \sqrt{9x - 15}$

Solve for x

$$5 + \sqrt{\tfrac{1}{2}x + 9} = 12$$

Find $f \circ g(x)$

$$f(x) = \sqrt{2x + x}$$
$$g(x) = 3x - 5$$

Simplify $\left(\frac{x^3}{x^5} \right)^5$

$(x^{3-5})^5$

$(x^{-2})^5$

x^{-10}

$\frac{1}{x^{10}}$

Solve for x $3^{(x-4)} = 243$

using logarithms $log3^{(x-4)} = log243$

$x - 4log3 = log243$

divide $x - 4 = \frac{log243}{log3}$

$x - 4 = 5$

$x = 5 + 4$

$x = 9$

Simplify $\left(\frac{x^3}{x^5} \right)^5$

Solve for x $3^{(x - 4)} = 243$

Find $f \circ g(x)$

$f(x) = 4x^2 + 2$
$g(x) = x + 5$

---

insert g(x) $4(x + 5)^2 + 2$

$4(x + 5)(x + 5) + 2$

using FOIL $4(x^2 + 10x + 25) + 2$

combine like terms $4x^2 + 40x + 100 + 2$

$f \circ g(x) = 4x^2 + 40x + 102$

Simplify

$-x(10 - 6 \div 2 \cdot 3 + 11)$

---

follow PEMDAS $-x(10 - 3 \cdot 3 + 11)$

$-x(10 - 9 + 11)$

$-x(1 + 11)$

$-x\ (12)$

$-12x$

Find $f \circ g(x)$

$$f(x) = 4x^2 + 2$$
$$g(x) = x + 5$$

Simplify

$$-x(10 - 6 \div 2 \cdot 3 + 11)$$

Solve for x $\sqrt{x\sqrt{49}} = x$

$\sqrt{7x} = x$

square both sides $(\sqrt{7x})^2 = x^2$

$7x = x^2$

set to equal '0' $7x - x^2 = 0$

factor $x(-x + 7) = 0$

$x = 0$ *or* $x = 7$

Solve for x $\frac{x}{7}(7) + x = 10$

$\frac{7x}{7} + x = 10$

multiply by 7 $7x + 7x = 70$

combine like terms $14x = 70$

divide $x = \frac{70}{14}$

$x = 5$

Solve for x $\sqrt{x\sqrt{49}} = x$

Solve for x $\frac{x}{7}(7) + x = 10$

Find $f \circ g(5)$

$f(x) = x^2 + 7$
$g(x) = \sqrt{x + 4}$

---

insert 5 to g(x) $\sqrt{5 + 4}$

insert g(5) to f(x) $(\sqrt{5 + 4})^2 + 7$

$(\sqrt{9})^2 + 7$

$(3)^2 + 7$

$9 + 7 = 16$

16

Find $g \circ f(5)$

$f(x) = x^2 + 7$
$g(x) = \sqrt{x + 4}$

---

insert 5 to f(x) $5^2 + 7$

insert f(5) to g(x) $\sqrt{(5^2 + 7) + 4}$

$\sqrt{25 + 7 + 4}$

$\sqrt{36}$

6

Find $f \circ g(5)$

$$f(x) = x^2 + 7$$
$$g(x) = \sqrt{x + 4}$$

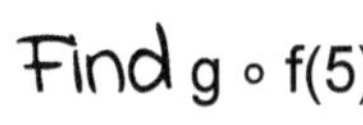

$g \circ f(5)$

$$f(x) = x^2 + 7$$
$$g(x) = \sqrt{x + 4}$$

Solve for x $\sqrt{\sqrt{81} + x} = x - 3$

$\sqrt{9 + x} = x - 3$

square both sides $(\sqrt{9 + x})^2 = (x - 3)^2$

using FOIL $9 + x = (x - 3)(x - 3)$

$9 + x = x^2 - 6x + 9$

set to equal '0' $9 + x - x^2 + 6x - 9 = 0$

$-x^2 + 7x = 0$

factor $x(-x + 7) = 0$

$x = 0$ *or* $x = 7$

Solve for x $(x^2)^3 \cdot x^2 = 256$

$x^6 \cdot x^2 = 256$

$x^8 = 256$

$x = 256^{\frac{1}{8}}$

$x = -2$ *or* $x = 2$

Solve for x $\sqrt{\sqrt{81} + x} = x - 3$

Solve for x $(x^2)^3 \cdot x^2 = 256$

Solve for x

$$3(x - 8) + \sqrt{x} = -x - 6$$

$$3x - 24 + \sqrt{x} = -x - 6$$

isolate $\sqrt{x}$

$$\sqrt{x} = -x - 6 - 3x + 24$$

combine like terms

$$\sqrt{x} = -4x + 18$$

square both sides

$$(\sqrt{x})^2 = (-4x + 18)^2$$

using FOIL

$$x = (-4x + 18)(-4x + 18)$$

$$x = 16x^2 - 144x + 324$$

set to equal '0'

$$x - 16x^2 + 144x - 324 = 0$$

combine like terms

$$-16x^2 + 145x - 324 = 0$$

factor

$$(-16x + 81)(x - 4)$$

$$x = 4$$

Simplify

$$\frac{(x^5)(x)}{x^3} \cdot \frac{x^8}{x^2}$$

$$\frac{(x^6)}{x^3} \cdot \frac{x^8}{x^2}$$

$$x^3 \cdot x^6$$

$$x^9$$

Solve for x $3(x - 8) + \sqrt{x} = -x - 6$

Simplify $\frac{(x^5)(x)}{x^3} \cdot \frac{x^8}{x^2}$

Find $f \circ g(x)$

$f(x) = x - 2x$
$g(x) = (x - 3)^2$

square g(x) $g(x) = (x - 3)(x - 3)$

using FOIL $g(x) = x^2 - 6x + 9$

insert g(x) $x^2 - 6x + 9 - 2(x^2 - 6x + 9)$

$x^2 - 6x + 9 - 2x^2 + 12x - 18$

combine like terms $-x^2 + 6x - 9$

$-x^2 + 6x - 9$

Simplify

$\sqrt{30}(\sqrt{5} + 2)$

$(\sqrt{30} \cdot \sqrt{5}) + (\sqrt{30} \cdot 2)$

$(\sqrt{150}) + (2\sqrt{30})$

factor $\sqrt{150}$ $(\sqrt{25} \cdot \sqrt{6}) + (2\sqrt{30})$

$(5\sqrt{6}) + (2\sqrt{30})$

$2\sqrt{30} + 5\sqrt{6}$

Find $f \circ g(x)$

$$f(x) = x - 2x$$
$$g(x) = (x - 3)^2$$

Simplify

$$\sqrt{30}(\sqrt{5} + 2)$$

Solve for x $$\frac{2x+2}{x+1} = \frac{3}{x-1} - 1$$

$$\frac{(2x+2)(x-1)}{x+1} = \frac{3(x+1)}{x-1} - 1(x+1)$$

$$\frac{(2x^2-2)}{x+1} = \frac{(3x+3)}{x-1} - 1x - 1$$

$$\frac{(2x^2-2)(x+1)}{x+1} = \frac{(3x+3)(x-1)}{x-1} - 1(x-1)(x+1)$$

$$2x^2 - 2 = 3x + 3 - 1(x^2 - 1)$$

$$2x^2 - 2 = 3x + 3 - x^2 + 1$$

combine like terms $$2x^2 - 2 = -x^2 + 3x + 4$$

set to equal '0' $$2x^2 - 2 + x^2 - 3x - 4 = 0$$

combine like terms $$3x^2 - 3x - 6 = 0$$

factor $$3(x + 1)(x - 2)$$

$$x = 2$$

Solve for x $$\frac{2x+2}{x+1} = \frac{3}{x-1} - 1$$

Solve equations

Example:	$2x - 3 + 3x = x + 5$
Rearrange	$2x + 3x - 3 = x + 5$
Combine like terms	$5x - 3 = x + 5$
Isolate variable by moving variable to left of equal sign and everything else to right of equal sign. Use inverse operations.	$5x - x = 5 + 3$
Combine like terms again.	$4x = 8$
Isolate variable again using inverse operations.	$x = \frac{8}{4}$ $x = 2$

Factoring

Find roots of expression $x^2 + 5x + 6 = (x + 2)(x + 3)$

using $x^2 + (m+n)x + mn$

Use FOIL to check: $(x + 2)(x + 3)$

$x^2 + 2x + 3x + 6$

Combining like terms: $x^2 + 5x + 6$

Working with negative numbers

Add/Subtract

Like signs - Add

```
   5      -5
  +3     +-3
   8      -8
```

Unlike signs - Subtract

```
   5      -5
  -3      +3
   2      -2
```

Multiply/Divide

Like signs - Positive

```
   5      -5
 x 3    x -3
  15      15
```

Unlike signs - Negative

```
   5      -5
 x-3     x 3
 -15     -15
```

Algebra rules for arithmetic

$$a(b + c) = ab + ac$$

$$a\left(\frac{b}{c}\right) = \frac{ab}{c}$$

$$\frac{a}{b} + \frac{c}{d} = \frac{ad + bc}{bd}$$

$$\frac{a + b}{c} = \frac{a}{c} + \frac{b}{x}$$

$$\frac{ac + bc}{c} = a + b$$

Exponents

$$a^m a^n = a^{m+n}$$

$$(a^m)^n = a^{mn}$$

$$(ab)^n = a^n b^n$$

$$a^{-n} = \frac{1}{a^n}$$

$$\left(\frac{a}{b}\right)^{-n} = \left(\frac{b}{a}\right)^n$$

$$\frac{a^n}{a^m} = a^{n-m}$$

$$a^0 = 1$$

Properties of Logarithms

Logarithm of a Product $\log_a(MN) = \log_a M + \log_a N$

Logarithm of a quotient $\log_a \frac{M}{N} = \log_a M - \log_a N$

Logarithm of a power $\log_a M^p = p \log_a M$

Compositions $\mathbf{f \circ g(x) = f(g(x))}$

Find $f \circ g(x)$

$f(x) = x + 4$
$g(x) = 2 - x$

replace the x in f(x) with g(x)

$f \circ g(x) = 2 - x + 4$
$= - x + 6$

Find $f \circ g(4)$

$f(x) = x^2 + 6$
$g(x) = \sqrt{x + 1}$

replace the x in g(x) with 4

$g(x) = \sqrt{4 + 1}$
$= \sqrt{5}$

replace the x in f(x) with $\sqrt{5}$

$f \circ g(8) = \sqrt{5}^2 + 6$
$= 11$

Solve for x

$$\frac{1}{\frac{1}{x}+3} = \frac{3}{9+\frac{3}{x}}$$

multiply by 'x'

$$\frac{1(x)}{(x)\frac{1}{x}+3(x)} = \frac{3(x)}{9(x)+(x)\frac{3}{x}}$$

cross-multiply

$$\frac{1x}{1+3x} = \frac{3x}{9x+3}$$

$$9x^2 + 3x = 9x^2 + 3x$$

set to equal '0'

$$9x^2 + 3x - 9x^2 - 3x = 0$$

combine like terms

$$0 = 0$$

All real numbers are solutions

Solve for x

$$5 + \sqrt{x} = 11 - \sqrt{x}$$

isolate both $\sqrt{x}$

$$\sqrt{x} + \sqrt{x} = 11 - 5$$

combine like terms

$$2\sqrt{x} = 6$$

divide by 2

$$\sqrt{x} = \frac{6}{2}$$

$$\sqrt{x} = 3$$

square both sides

$$(\sqrt{x})^2 = 3^2$$

$$x = 9$$

Solve for x

$$\frac{1}{\frac{1}{x}+3} = \frac{3}{9+\frac{3}{x}}$$

Solve for x

$$5 + \sqrt{x} = 11 - \sqrt{x}$$

Solve for x	$x - \sqrt{x} = 2x - 6$
isolate $-\sqrt{x}$	$-\sqrt{x} = 2x - 6 - x$
combine like terms	$-1\sqrt{x} = x - 6$
divide by -1	$\sqrt{x} = \frac{x-6}{-1}$
	$\sqrt{x} = -x + 6$
square both sides	$(\sqrt{x})^2 = (-x + 6)^2$
using FOIL	$(\sqrt{x})^2 = (-x + 6)(-x + 6)$
	$x = x^2 - 12x + 36$
set to equal '0'	$-x^2 + 12x + x - 36 = 0$
combine like terms	$-x^2 + 13x - 36 = 0$
factor	$(-x + 4)(x - 9) = 0$
	$x = 4$

Solve for x $x - \sqrt{x} = 2x - 6$

Solve for x $\frac{1}{5}x = -x + 18\left(\frac{1}{3}\right)$

$\frac{1x}{5} = -x + \frac{18}{3}$

multiply by 5 $\frac{1x}{5} = -x + 6$

$1x = -5x + 30$

combine like terms $1x + 5x = 30$

$6x = 30$

$x = \frac{30}{6}$

$x = 5$

Simplify $\sqrt{5x^3y} \cdot 2\sqrt{x}$

multiply $2\sqrt{5x^4y}$

simplify $2x^2\sqrt{5y}$

Solve for x $\frac{1}{5}x = -x + 18\left(\frac{1}{3}\right)$

Simplify $\sqrt{5x^3y} \cdot 2\sqrt{x}$

Solve for y $x = \frac{2(3-y)}{5y}$

$x \cdot 5y = 2(3 - y)$

$5xy = 6 - 2y$

$5xy + 2y = 6$

factor $y(5x + 2) = 6$

$y = \frac{6}{5x+2}$

Solve for b $x = 1 - ab(\frac{1}{3})$

$x = 1 - \frac{ab}{3}$

$x - 1 = -\frac{ab}{3}$

multiply by 3 $3x - 1 = -ab$

$\frac{3x-1}{-a} = b$

$b = -\frac{3x-1}{a}$

Solve for y $x = \frac{2(3 - y)}{5y}$

Solve for b $x = 1 - ab(\frac{1}{3})$

Solve for x $x\sqrt{x} - 7 = x + 11$

isolate radical $x\sqrt{x} = x + 11 + 7$

combine like terms $x\sqrt{x} = x + 18$

square both sides $(x\sqrt{x})^2 = (x + 18)^2$

using FOIL $(x\sqrt{x})(x\sqrt{x}) = (x + 18)(x + 18)$

$x^3 = x^2 + 36x + 324$

$x^3 - x^2 - 36x - 324 = 0$

factor $(x - 9)(x^2 + 8x + 36) = 0$

$x = 9$

Solve for x $2 - (x - 4)(3) = 1 - (-x - 1)$

$2 - 3x + 12 = 1 + x + 1$

combine like terms $-3x + 14 = x + 2$

$-3x - x = 2 - 14$

combine like terms $-4x = -12$

$x = \frac{-12}{-4}$

$x = 3$

Solve for x $x\sqrt{x} - 7 = x + 11$

Solve for x $2 - (x - 4)(3) = 1 - (-x - 1)$

Solve for y $-1 = 1 - \frac{y-7}{y+7}$

isolate the fraction $-1 - 1 = -\frac{y-7}{y+7}$

combine like terms $-2 = \frac{-y+7}{y+7}$

multiply by y + 7 $-2y - 14 = -y + 7$

combine like terms $-2y + y = 7 + 14$

$-1y = 21$

divide $y = \frac{21}{-1}$

$y = -21$

Combine terms $3y^2(x - y) - (-y^2 + x) + 2x(y - 1)$

multiply $3xy^2 - 3y^3 - (y^2 + x) + 2xy - 2x$

rewrite $3xy^2 - 3y^3 - y^2 - x + 2xy - 2x$

rewrite $3xy^2 - 3y^3 + 2xy - y^2 - 2x - x$

combine like terms $3xy^2 - 3y^3 + 2xy - y^2 - 3x$

Solve for y

$$-1 = 1 - \frac{y-7}{y+7}$$

Combine terms

$$3y^2(x - y) - (-y^2 + x) + 2x(y - 1)$$

Solve for x $-(x - 2)(x + 3) = 2x - 4$

using Foil $(-x + 2)(x + 3) = 2x - 4$

$-x^2 - x + 6 = 2x - 4$

set to equal '0' $-x^2 - x + 6 - 2x + 4 = 0$

combine like terms $-x^2 - 3x + 10 = 0$

factor $(-x + 2)(x + 5) = 0$

$x = -5$ *or* $x = 2$

Solve for x $3 - x \div 2 \cdot 5 + 1 = -6$

subtract the 3 & 1 $-x \div 2 \cdot 5 = -6 - 4$

$-x \div 2 \cdot 5 = -10$

rewrite $\frac{-x}{2} \cdot 5 = -10$

multiply 5 to -x $\frac{-5x}{2} = -10$

multiply 2 to -10 $-5x = -10 \cdot 2$

$-5x = -20$

divide $x = \frac{-20}{-5}$

$x = 4$

Solve for x $-(x - 2)(x + 3) = 2x - 4$

Solve for x $3 - x \div 2 \cdot 5 + 1 = -6$

Find $f \circ g(5)$

$$f(x) = (x^2)$$
$$g(x) = \frac{x}{y} + 3$$

insert 5 to g(x)

$$\frac{5}{y} + 3$$

insert g(x) to f(x)

$$(\frac{5}{y} + 3)^2$$

$$(3\frac{5}{y})(3\frac{5}{y})$$

using FOIL

$$(\frac{3y+5}{y})(\frac{3y+5}{y})$$

$$\frac{9y^2 + 30y + 25}{y^2}$$

Find $f \circ g(3)$

$$f(x) = (-x - 2y)^2$$
$$g(x) = -x - 2$$

insert 3 to g(x)

$$-3 - 2$$

insert g(x) to f(x)

$$(--3 - 2 - 2y)^2$$

combine like terms

$$(1 - 2y)^2$$

using FOIL

$$(1 - 2y)(1 - 2y)$$

$$4y^2 - 4y + 1$$

Find $f \circ g(5)$

$$f(x) = (x^2)$$
$$g(x) = \frac{x}{y} + 3$$

Find $f \circ g(3)$

$$f(x) = (-x - 2y)^2$$
$$g(x) = -x - 2$$

Solve for x $1 - x \cdot 4 \div 2 = -13$

subtract 1 from -13 $-x \cdot 4 \div 2 = -13 - 1$

$-x \cdot 4 \div 2 = -14$

rewrite $\frac{-4x}{2} = -14$

isolate -4x $-4x = -14 \cdot 2$

$-4x = -28$

isolate x $x = \frac{-28}{-4}$

$x = 7$

Solve for x $4(x - 2) \div 2 = 12$

multiply and rewrite $\frac{4x - 8}{2} = 12$

$4x - 8 = 12 \cdot 2$

isolate 4x $4x - 8 = 24$

$4x = 24 + 8$

$4x = 32$

isolate x $x = \frac{32}{4}$

$x = 8$

Solve for x

$$1 - x \cdot 4 \div 2 = -13$$

Solve for x

$$4(x - 2) \div 2 = 12$$

Simplify $\dfrac{12}{7-\sqrt{43}}$

multiply by the conjugate of the denominator $\dfrac{12}{7-\sqrt{43}} \cdot \dfrac{7+\sqrt{43}}{7+\sqrt{43}}$

$$\frac{84 + 12\sqrt{43}}{49 + 7\sqrt{43} - 7\sqrt{43} - 43}$$

combine like terms $\dfrac{84 + 12\sqrt{43}}{49 - 43}$

$$\frac{84 + 12\sqrt{43}}{6}$$

simplify' $\dfrac{84}{6} + \dfrac{12\sqrt{43}}{6}$

$$\frac{14}{1} + \frac{2\sqrt{43}}{1}$$

$$14 + 2\sqrt{43}$$

Simplify

$$\frac{12}{7-\sqrt{43}}$$

Factor	$6ab - 36ax + 5b - 30x$
find numbers	$6 \cdot 6 = 36$ $6 \cdot 5 = 30$
establish variables	a, b, and x
construct	$(6a + 5)(b - 6x)$
test with FOIL	$6ab - 36ax + 5b - 30x$
	$(6a + 5)(b - 6x)$

Find $f(x + 6)$	$f(x) = 2x^2 + 4$
replace x with x + 6	$2(x + 6)^2 + 4$
	$2(x + 6)(x + 6) + 4$
using FOIL	$2(x^2 + 12x + 36) + 4$
	$2x^2 + 24x + 72 + 4$
	$2x^2 + 24x + 76$

Factor $6ab - 36ax + 5b - 30x$

Find $f(x + 6)$ $f(x) = 2x^2 + 4$

Find f(x - 3) $f(x) = \frac{x}{\sqrt{x} + 4}$

replace x with x - 3

$$\frac{x - 3}{\sqrt{x - 3} + 4}$$

multiply by the conjugate of the denominator

$$\frac{x - 3}{\sqrt{x - 3} + 4} \cdot \frac{\sqrt{x - 3} - 4}{\sqrt{x - 3} - 4}$$

$$\frac{(x - 3)\sqrt{x - 3} - 4x + 12}{(\sqrt{x - 3})^2 - 4\sqrt{x - 3} + 4\sqrt{x - 3} - 16}$$

combine like terms

$$\frac{(x - 3)\sqrt{x - 3} - 4x + 12}{(\sqrt{x - 3})^2 - 16}$$

squaring the radical

$$\frac{(x - 3)\sqrt{x - 3} - 4x + 12}{x - 3 - 16}$$

$$\frac{(x - 3)\sqrt{x - 3} - 4x + 12}{x - 19}$$

Find f(x - 3) $f(x) = \frac{x}{\sqrt{x} + 4}$

Find g(-x + 5) $g(x) = \dfrac{x}{7 - \sqrt{x}}$

replace x with -x + 5

$$\frac{-x+5}{7-\sqrt{-x+5}}$$

multiply by the conjugate of the denominator

$$\frac{-x+5}{7-\sqrt{-x+5}} \cdot \frac{7+\sqrt{-x+5}}{7+\sqrt{-x+5}}$$

$$\frac{(-x+5)\sqrt{-x+5}-7x+35}{49+7\sqrt{-x+5}-7\sqrt{-x+5}-(\sqrt{-x+5})^2}$$

combine like terms

$$\frac{(-x+5)\sqrt{-x+5}-7x+35}{49-(\sqrt{-x+5})^2}$$

squaring the radical

$$\frac{(-x+5)\sqrt{-x+5}-7x+35}{49+x-5}$$

$$\frac{(-x+5)\sqrt{-x+5}-7x+35}{x+44}$$

Find g(-x + 5) $\qquad$ $g(x) = \dfrac{x}{7 - \sqrt{x}}$

Find $f(-x - 2)$	$f(x) = 2x^2 - 3x + 5$
replace x with -x - 2	$2(-x - 2)^2 - 3(-x - 2) + 5$
	$2(-x - 2)(-x - 2) - 3(-x - 2) + 5$
using FOIL	$2(x^2 + 4x + 4) - 3(-x - 2) + 5$
	$2x^2 + 8x + 8 + 3x + 6 + 5$
combine like terms	$2x^2 + 11x + 19$

Find $g(x + 1)$	$g(x) = x(x - 4) + 2$
replace x with x + 1	$(x + 1)(x + 1 - 4) + 2$
	$(x + 1)(x - 3) + 2$
using FOIL	$(x^2 - 2x - 3) + 2$
combine like terms	$x^2 - 2x - 3 + 2$
	$x^2 - 2x - 1$

Find $f(-x - 2)$ $\quad f(x) = 2x^2 - 3x + 5$

Find $g(x + 1)$ $\quad g(x) = x(x - 4) + 2$

Simplify

$$\frac{\frac{1}{x+3}+4}{\frac{1}{x-3}-2}$$

multiply by (x + 3) and (x - 3)

$$\frac{(x+3)(x-3)\frac{1}{x+3}+4(x+3)(x-3)}{(x+3)(x-3)\frac{1}{x-3}-2(x+3)(x-3)}$$

$$\frac{\frac{1(x+3)(x-3)}{x+3}+4(x+3)(x-3)}{\frac{1(x+3)(x-3)}{x-3}-2(x+3)(x-3)}$$

cancel out

$$\frac{1(x-3)}{1(x+3)} \quad \frac{\frac{(x+3)}{x+3}}{\frac{(x-3)}{x-3}} \quad \frac{+4(x+3)(x-3)}{-2(x+3)(x-3)}$$

using FOIL

$$\frac{1x-3+4(x^2-3x+3x-9)}{1x+3-2(x^2-3x+3x-9)}$$

$$\frac{1x-3+4x^2-36}{1x+3-2x^2+18}$$

combine like terms

$$\frac{4x^2+x-39}{-2x^2+x+21}$$

Simplify

$$\frac{\frac{1}{x+3}+4}{\frac{1}{x-3}-2}$$

Simplify

$$\frac{\frac{2}{-x-5}-2}{-3+\frac{3}{x+5}}$$

multiply by (-x - 5) and (x + 5)

$$\frac{(-x-5)(x+5)\frac{2}{-x-5}-2(-x-5)(x+5)}{(-x-5)(x+5)\frac{3}{x+5}-3(-x-5)(x+5)}$$

$$\frac{\frac{2(-x-5)(x+5)}{-x-5}-2(-x-5)(x+5)}{\frac{3(-x-5)(x+5)}{x+5}-3(-x-5)(x+5)}$$

cancel out

$$\frac{2(x+5)}{3(-x-5)} \quad \frac{\frac{(-x-5)}{-x-5}}{\frac{(x+5)}{x+5}} \quad \frac{-2(-x-5)(x+5)}{-3(-x-5)(x+5)}$$

using FOIL

$$\frac{2x+10-2(-x^2-5x-5x-25)}{-3x-15-3(-x^2-5x-5x-25)}$$

$$\frac{2x+10+2x^2+20x+50}{-3x-15+3x^2+30x+75}$$

combine like terms

$$\frac{2x^2+22x+60}{3x^2+27x+60}$$

factor

$$\frac{2(x+6)(x+5)}{3(x+4)(x+5)}$$

cancel out

$$\frac{2(x+6)}{3(x+4)} \quad \frac{(x+5)}{(x+5)}$$

$$\frac{2x+12}{3x+12}$$

Simplify

$$\frac{\frac{2}{-x-5}-2}{-3+\frac{3}{x+5}}$$

Solve for x

$$\frac{\frac{2}{x+3}+2}{\frac{3}{x-3}+4} = 1$$

multiply by (x + 3) and (x - 3)

$$\frac{(x+3)(x-3)\frac{2}{x+3}+2(x+3)(x-3)}{(x+3)(x-3)\frac{3}{x-3}+4(x+3)(x-3)} = 1$$

$$\frac{\frac{2(x+3)(x-3)}{x+3}+2(x+3)(x-3)}{\frac{3(x+3)(x-3)}{x-3}+4(x+3)(x-3)} = 1$$

cancel out

$$\frac{2(x-3)}{3(x+3)} \quad \frac{\frac{(x+3)}{x+3}}{\frac{(x-3)}{x-3}} \quad \frac{+2(x+3)(x-3)}{+4(x+3)(x-3)} = 1$$

using FOIL

$$\frac{2x-6+2(x^2-3x+3x-9)}{3x+9+4(x^2-3x+3x-9)} = 1$$

$$\frac{2x-6+2x^2-18}{3x+9+4x^2-36} = 1$$

combine like terms

$$\frac{2x^2+2x-24}{4x^2+3x-27} = 1$$

multiply

$$2x^2 + 2x - 24 = 1(4x^2 + 3x - 27)$$

$$2x^2 + 2x - 24 = 4x^2 + 3x - 27$$

set to equal '0'

$$2x^2 + 2x - 24 - 4x^2 - 3x + 27 = 0$$

combine like terms

$$-2x^2 - 1x + 3 = 0$$

factor

$$(-2x - 3)(x - 1)$$

$$x = \frac{-3}{2} \quad \text{or} \quad x = 1$$

Solve for x

$$\frac{\frac{2}{x+3}+2}{\frac{3}{x-3}+4} = 1$$

On the cover

Solve for x $$1 - \frac{x-1}{\sqrt{x+1}} = 9x$$

simplify the fraction $$1 - \frac{(x-1)\sqrt{x+1}}{x+1} = 9x$$

multiply by x + 1 $$1x + 1 - (x - 1)\sqrt{x+1} = 9x^2 + 9x$$

isolate radical $$-(x - 1)\sqrt{x+1} = 9x^2 + 9x - 1x - 1$$

$$(-x + 1)\sqrt{x+1} = 9x^2 + 8x - 1$$

$$\sqrt{x+1} = \frac{9x^2 + 8x - 1}{-x+1}$$

square both sides $$(\sqrt{x+1})^2 = \left(\frac{9x^2 + 8x - 1}{-x+1}\right)^2$$

$$x + 1 = \frac{81x^4 + 144x^3 + 46x^2 - 16x + 1}{x^2 - 2x + 1}$$

$$(x + 1)(x^2 - 2x + 1) = 81x^4 + 144x^3 + 46x^2 - 16x + 1$$

$$x^3 - x^2 - x + 1 = 81x^4 + 144x^3 + 46x^2 - 16x + 1$$

$$-81x^4 - 143x^3 - 47x^2 + 15x = 0$$

factor $$x(-x - 1)(81x^2 + 62x - 15) = 0$$

$x = 0$ or $x = -1$ or $x = 0.19318$ or $x = -0.95861$

only x = 0.19318 works in original equation

Solve for x

$$1 - \frac{x-1}{\sqrt{x+1}} = 9x$$

Algebra Definitions

Expressions A mathematical phrase that may contain numbers, variables, or operations, but does not include a relationship symbol such as =, <, or >.

Variables A letter used that represents a quantity that can change.

Coefficient The number part of a term with a variable.

Constants A term containing only numbers. Constants do not include variables.

Binomial A polynomial with two terms.

Polynomial An expression consisting of variables and coefficients that involves only the operations of addition, subtraction, multiplication, and non-negative integer exponents.

Inequality A sentence that states one expression is greater than, less than or equal to another expression

Conjugate A math conjugate is formed by changing the sign between two terms in a binomial.

Other books from Timothy Schablin Mathematics

equals(me)
Pre-Algebra Practice

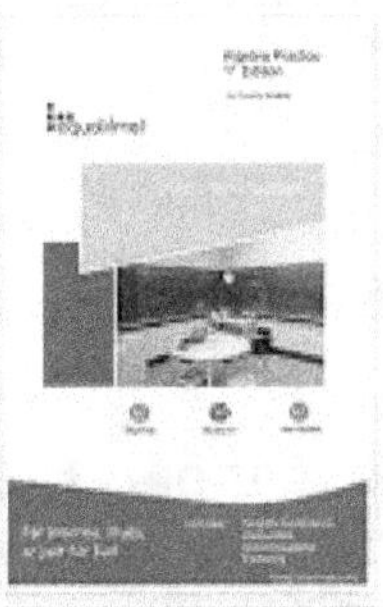

equals(me)
Algebra Practice

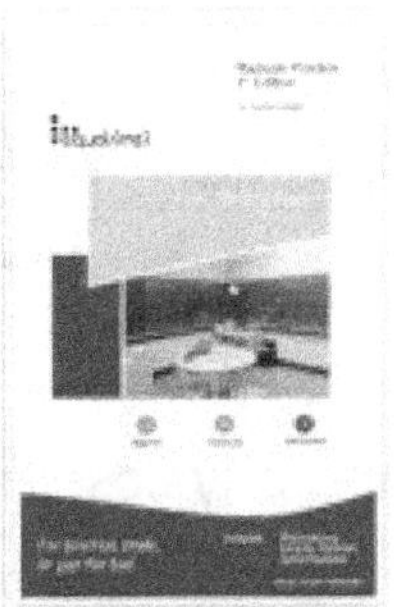

equals(me)
Radicals Practice

Fun with PEMDAS
PEMDAS Practice

Available at Amazon Books or

https://timothyschablin.wixsite.com/equalsme

About the author

Timothy Schablin is a graduate of the Hutchinson Technical College where he studied algebra, trigonometry, physics, mathematical techniques, and technical related fields. He also holds two certificates of physics from Davidson College, AP Physics I & AP Physics II: Challenging Concepts.

Timothy Schablin tutors math to 5th, 6th, 7th, and 8th graders at a local middle school. He is also a member of Minnesota MathCorps and has authored mathematical software.

Besides studying & tutoring math and physics, Timothy enjoys astronomy. He spends vacation time canoeing the Minnesota River bottom.

Copyright © 2018 Timothy Schablin

www.ingramcontent.com/pod-product-compliance
Lightning Source LLC
Chambersburg PA
CBHW072015230526
45468CB00021B/1564

* 9 7 8 1 7 2 0 9 0 0 3 7 5 *